Where's Polygon?

By Patricia Akroyd

Illustrations by Sarah McMurray

Spiritual "equipment" for the contest of life.
www.Spiritbuilding.com

SPIRITBUILDING PUBLISHING
15591 N. State Rd. 9, Summitville, Indiana, 46070

ISBN 978-0-9829811-8-4

© SPIRITBUILDING, 2014, All Rights Reserved. No part of this book may be reproduced in any form without the written permission of the publisher. Printed in the United States of America.

This book is designed to teach geometry. It's meant as an introduction to the polygon shapes. I recommend that the book should first be merely read to the students. Next, it should be reread, sharing the illustrations and discussing the attributes of each kind of polygon. I think that it would be a good exercise for the students to make their own illustrations of the sections or characters. They could even do some creative writing based on the characters, expanding the story.

Patricia Akroyd

Before real time began, there was Geometry World.

And in Geometry World, the kingdom of Queen Polygon.
In this kingdom, all shapes lived and worked s e p a r a t e l y .

Queen Polygon was the ruler in the land of Polygon.
All her subjects were **closed, flat,** polygon shapes with **straight line** borders, **angles** and **corners** (corners are called vertices).

If any line crossed or intersected another, that shape lost its citizenship!

The **Polygons** lived in separate villages, which they called **Sections**, and were suspicious of any shape with **different** types of line segments. They only trusted their Queen.

Queen Poly, as she was best known, could be the shape of **any** of the Polygons, but each group saw her **only** as their own shape.

The lowest possible number in Geometry world was **three**,
so that is where Queen Poly's realm began.

Section Three was called **Tri Village**.

Citizens in Tri Village were called **Triangles**, or **Tris** for short.
Everything in their lives was in threes,
because "tri" is the prefix meaning three.

For example, a **tricycle** has **three** wheels.

The Triangles had three sides, which were closed, **three angles** and three vertices.

Even within Tri Village, the Triangles were s e p a r a t e d into groups. The most elite Tris were the **Right Triangles**.

If a Tri were a Right, it had an angle like an "L," which is a right angle, having two lines **perpendicular** to each other.

This characteristic gave them the reputation of being up**right**. Naturally, the other Triangles thought that someone who looked up**right** should be trusted to make the major decisions for the village.

Therefore, the **Right Triangles** were the ones chosen to govern the village.

Isosceles Trianges, having TWO equal sides, were the teachers of the village, because the name sounded so philosophical.
It is even rumored that in Real Time, they insprired the Egyptian **pyramids.**

The common citizens were triangles with all **different** sides and angles.

They are the **Scalene Triangles.**
There was an occasional "black sheep" in this group, which had an **obtuse** angle.
Even with such an unusual angle though, the Triangles still considered them family.

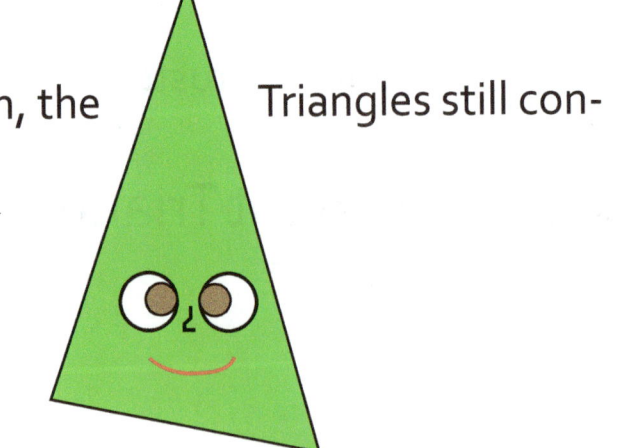

The next village was the largest.
In fact, you could call it a City.

Section **Four** was **Quad** City. That's because **quad** means **four**.
Everyting in the City was in groups of four.

Its citizens were called **Quadrilaterals**.

What they had in common was **four** line segments, **four** angles and **four** vertices (Remember, vertices are corners!).
All *vertices* added up to 360 degrees, no more and no less.

Like the other villages, the **Quadrilaterals** were also divided into classes.

The largest group of citizens was the para*ll*elograms. Para*ll*elograms can have **four right** angles, or **two acute** angles and **two obtuse** angles. No matter what, they have two sets of para*ll*el line segments. That's how they got their name!

The upperclass were the para*ll*elograms called rectangles. Rectangles were the ones with **four right** angles. Even rectangles were divided into groups. There were the o b l o n g rectangles and the squares.

Both o b l o n g rectangles and squares make their sides from perpendicular lines with **right** angles.

This looks like an L.

The elite even among the rectangles were the squares. These are rectangles that also have **four** EQUAL sides. Because of all this "right-ness" and EQUALity, the squares were most trusted to make decisions for Quad City.

Occasionally, a *square* would bend and become a *rhombus*.

That happened when **two** of the **opposite sides** would slide in **opposite directions**. The Right angles would then change into **two acute** and **two obtuse** angles, but still with **four** EQUAL sides!

Although this caused some embarrassment to the *square*, it was usually overlooked because a *rhombus* is also a *diamond*!

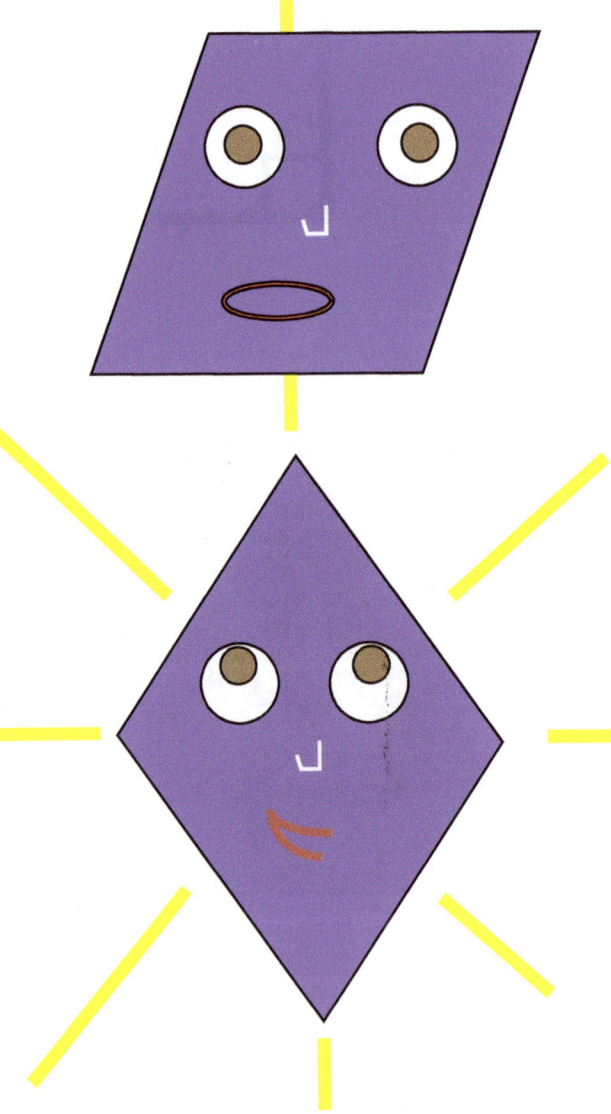

In the poor section of Quad City lived the traps. Their full name was Trapezoid. Of course, they had the required four sides, but those sides could be all different lengths.

Their angles could be of any type: **acute**, **obtuse**, or **right**.

They were very agreeable fellows, but other Quads sometimes looked down their nose at them, since they were unique.

Still, life in Quad City was serene and productive.

One of the nearby villages was Penta Village, which was located in Section Five. Pentagons were the citizens of Penta Village, because "penta" means five, and Pentagons have five sides.

They had it all: three types of angles (**acute**, **obtuse**, and **right**)!

What more could a Penta want?

Their most common task was forming the shape of a HOUSE.

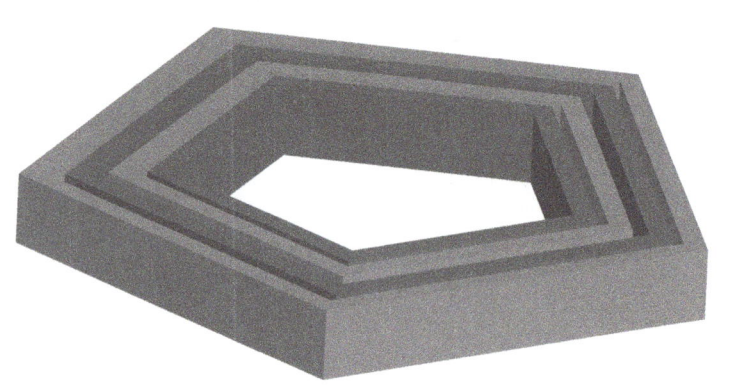

They were especially proud of the fact that in *Real Time* a famous building in Washington D.C. was named after them.

Just down the road in Section **Six** was **Hexa** Town, home of the **Hexagons**. With their **six** sides and **six** angles (mostly **obtuse**, but with the occasional **right**), the **Hexas** didn't envy any of the other villages.

They were too busy with the b e e s . Their main occupation was forming the cells of b e e hives. And I'm sure you have heard how busy b e e s are!

When **Hexas** weren't tending b e e hives, they were making floor tiles, or fancy party crackers.

Although it was rare for the different villages to get together, the **Hexagons** would occasionally join their neighbors, the **Pentagons**, on a *soccer ball* for a day out.

In Section **Seven** was the **Heptagons** of **Hepta** Town. **"Hepta"** means **seven**.

Few could figure out exactly what their main occupation was, but they always seemed to be in *SEVENTH HEAVEN*.

Unlike the placid **Heptagons**, the next village, in Section **Eight**, housed the **Octagons**. Their town was called **Octa** Villa.

The **Octas** were always yelling **"STOP,"** since they were shaped like an **eight**-sided **stop** sign.

Of course they did other things too, but **"STOP"** was what they were best known for.

With their **eight** sides, **eight** angles and **eight** vertices, they were a very loud group.

On the surface, Queen Poly's kingdom seemed fairly peaceful. There were no fights, because violence was unknown in GEOMETRY WORLD.

Nevertheless, Queen Poly was concerned.

The Polygons would not communicate with each other. They kept mostly to themselves and were very *suspicious* of the other shapes. The tension was thick.

Queen **Poly** loved all her subjects and knew they were all related. She called a conference and required each village to send a REPRESENTATIVE shape.

The **polygons** attending were very leery of each other. They only trusted **Queen Poly,** because they saw her as their *own* shape. You recall that she could be any shape, as long as it was made of closed-line segments.

The conference meeting was a disaster. Nobody would communicate. The **Octa** representative kept yelling **"STOP!"** The **Hepta** representative seemed to be in a world all his own. The **Tri** was the most uncomfortable, because she had the fewest number of sides.

What was Poly to do? She felt quite perplexed and hopeless. She adjourned the meeting. But, being a persistent queen, she commanded them all to return in THREE days. Since THREE was the lowest number in the kingdom, that was the soonest possible day!

The only Polygon receptive to this command was the Triangle, since THREE was her number. This made her feel a little more important.

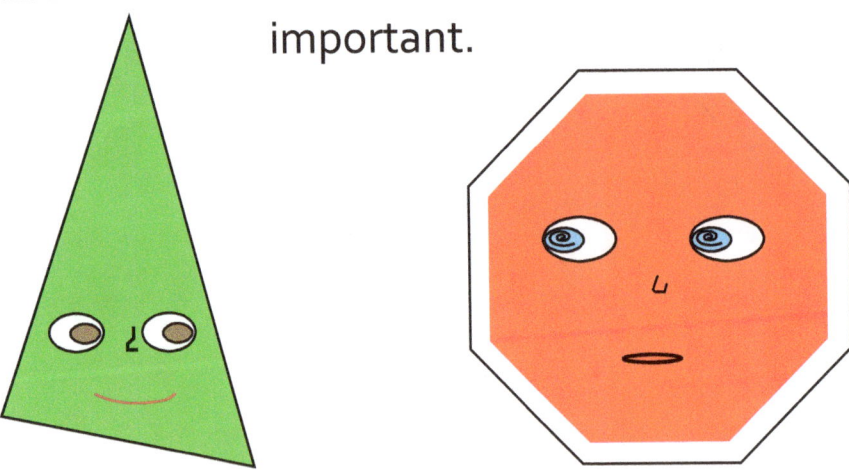

Queen Poly retired to her royal bed chamber and pondered the situation for hours. Having come to no solution, she finally fell asleep.

That night, EMPEROR GEO, ruler of all the kingdoms of GEOMETRY WORLD, came to Poly in a dream.

"Poly," he called to her, "you must do something *radical*!"

Poly!

"What should I do?" asked Queen Poly, "The Polygons will not communicate!"

?

"You are wise and creative, Queen Poly. You will figure it out. Just put your mind at ease and think on *positive thoughts*."

The next morning Queen Poly was inspired with a plan, just as Emperor Geo had promised.

Queen Poly formed her shape into a Triangle with THREE sides.

Then she became a Quadrilateral with **four** sides.

Next, she formed herself into a Pentagon with **five** sides.

She added another side and became a Hexagon with **six** sides.

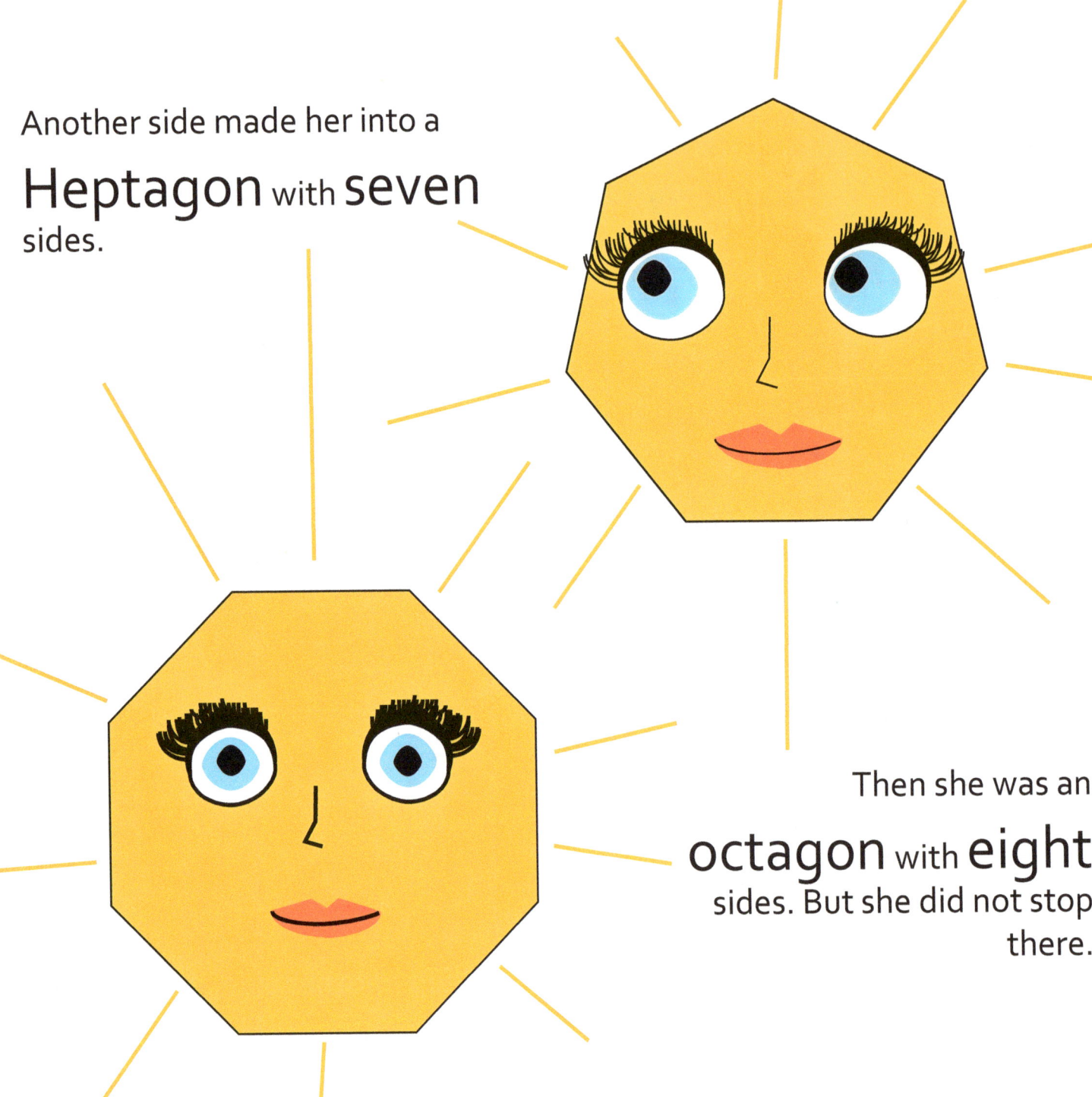

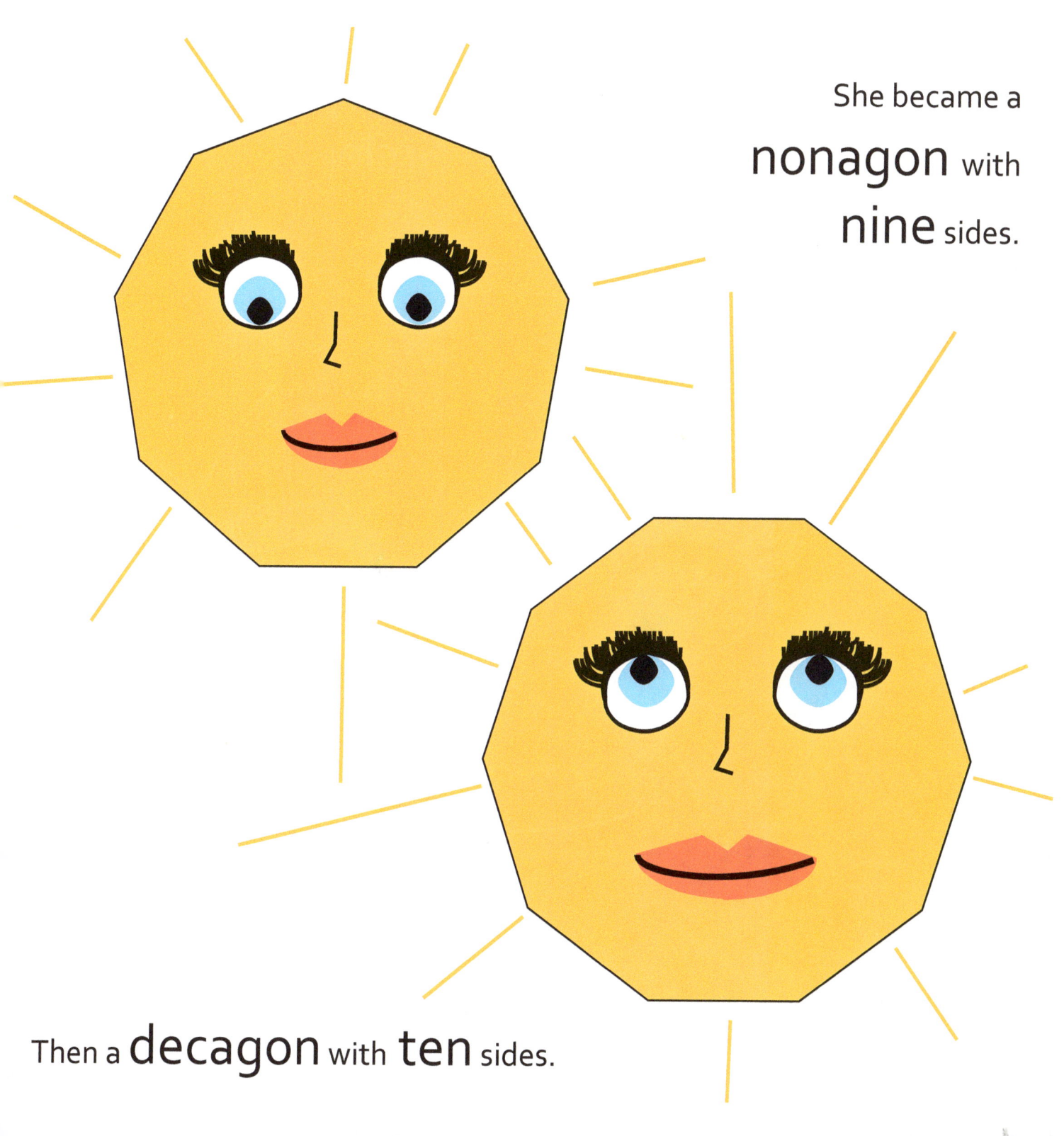

She became a **nonagon** with **nine** sides.

Then a **decagon** with **ten** sides.

On and on she kept adding sides, until there were so m a n y, and they were so SMALL, that she looked like a circle!

On the third day, the reluctant **polygon** representatives returned and entered the conference room.

They were astonished! The room seemed to have lost its **sides**. In the middle of the room was what looked like a CIRCLE!

"WHERE HAS POLY GONE?" they all exclaimed.

Actually, it was not exactly a CIRCLE. Queen Poly had added so many **sides** to herself, and each **side** had become so SMALL, that you would need a microscope to see them.

Her voice filled the room: "Be seated around me. You can no longer be isolated. We are going to face each other and settle our d*i*f f e r e n C e s once and for all!"

Queen **Poly** began by speaking to the **Hexagon**, "Now **Hexa**, we'll start with you."

"If you show a line of *symmetry*, that is, divide yourself EXACTLY in HALF, what will you be?"

Magically, a line divided the **Hexagon** into two **Pentagons**.

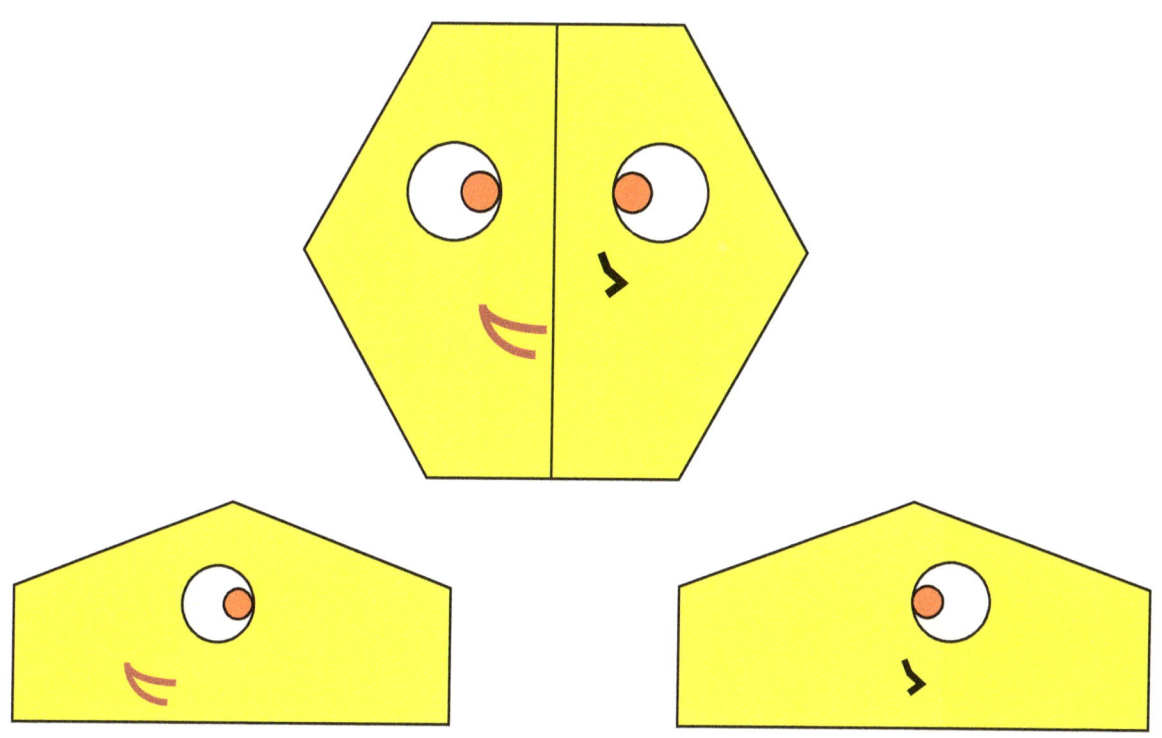

Everyone gasped in *amazement!*

Queen **Poly** continued, "What happens when you divide yourself in the other direction?"

The crowd was even more *amazed*.

"That's right! You are now two **Quadrilateral Trapezoids**."

"But what would happen if you connected all opposite **vertices** with *diagonal* lines? What are you made of now?"

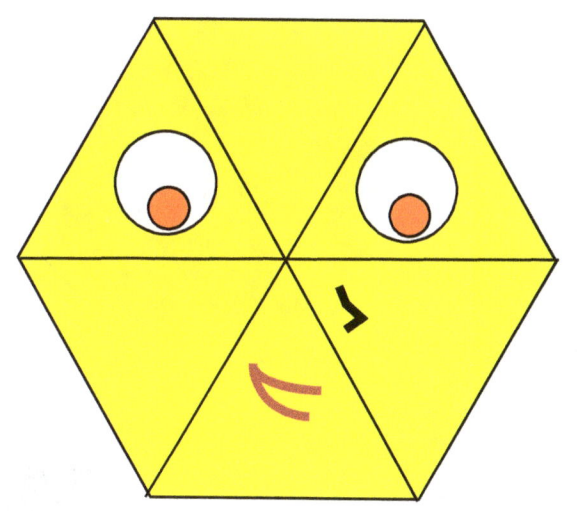

"Yes, you are made of **six Triangles!**"

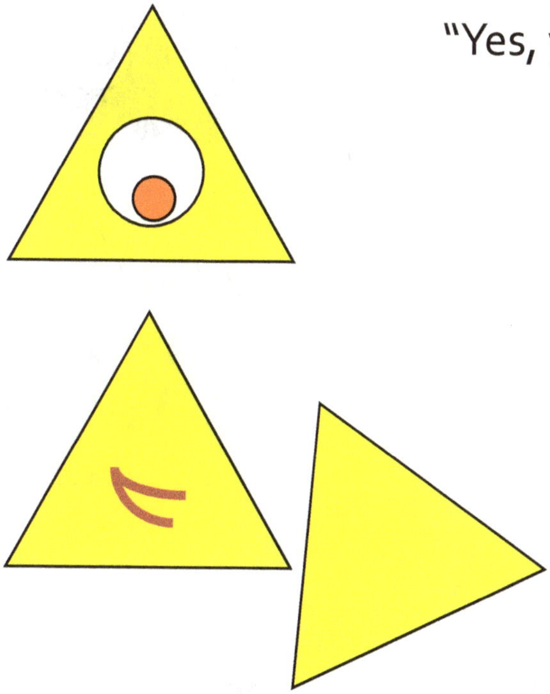

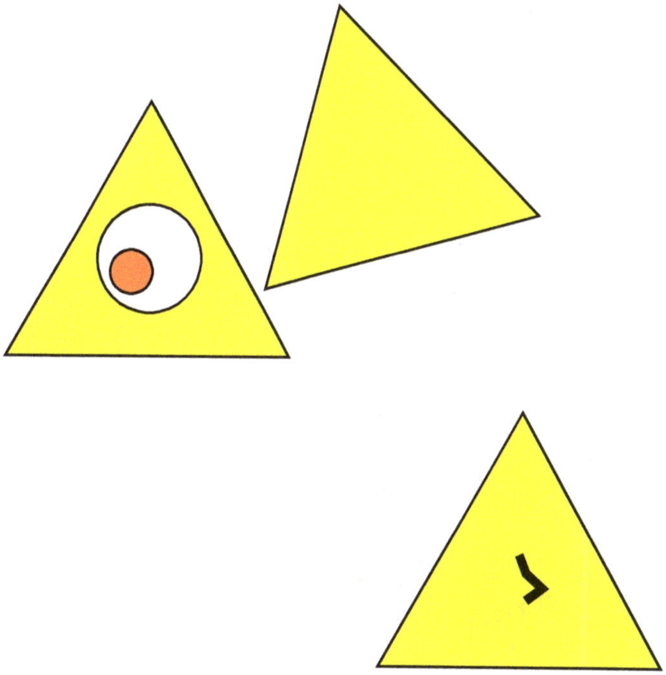

Queen **Poly** made all the **polygon** representatives begin drawing **LINES** within themselves.

They discovered that they were all made up of parts of each other.

Their heads filled with the possibilities. The polygon representatives all rushed back to their villages to share this new knowledge. They were now united POLYGONS!!!

The Polygons joined together and even invited the curved shapes to join them from other kingdoms. They were able to form all kinds of things to create the REAL WORLD, where we live.

POLY HAD GONE AND DONE IT!!!

The End.

Where's Polygon?

Glossary of Terms

Vertex: The point at which two lines intersect, to form an angle or corner.

Vertices: The plural of vertex.

Right angle: Two line segments that intersect 90 degrees from each other, making an 'L' shape. Also see perpendicular lines.

Acute angle: Two line segments that intersect to form an angle less than 90 degrees.

Obtuse angle: Two line segments that intersect to form an angle greater than 90 degrees.

Parallel lines: Two lines, equal distances apart. If you extended their segments infinitely into space, the two lines would never intersect.

Perpendicular lines: Two lines that intersect to form a right angle. See also right angle.

Line of symmetry: Any line drawn which divides a shape perfectly in half. If the shape were to be folded along the line, each side would match.

Equilateral: A polygon that has all segments of the same length.

Polygon: Any simple closed figure made of line segments.

Triangle: A polygon with three line segments and three angles (vertices). Each angle will add together to equal 180 degrees.

Right triangle: A triangle with one right angle (see right angle) and two acute angles (see acute angles). The two sides forming the right angle will be perpendicular (see perpendicular).

Equilateral triangle: A triangle with three equal sides and three equal angles. Each angle measures 60 degrees.

Isosceles triangle: A triangle with three sides, two of which are equal. Two of the three angles are also equal.

Quadrilateral: A polygon with four sides and four angles. The four angles add up to a total of 360 degrees.

Parallelogram: A quadrilateral with two sets of parallel lines.

Rectangle: A parallelogram with four right angles.

Square: A rectangle with four equal sides.

Rhombus: A parallelogram with four equal sides, two acute angles (see acute angles) and two obtuse angles (see obtuse angle).

Trapezoid: A quadrilateral with one set of parallel lines.

Pentagon: A polygon with five sides.

Hexagon: A polygon with six sides.

Heptagon: A polygon with seven sides.

Octagon: A polygon with eight sides.

Nonagon: A polygon with nine sides

Decagon: A polygon with ten sides.

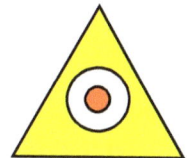

For Kid's Only

Greta's Purpose: A children's book about a Great Dane who struggles with fitting in.

Rudy's Path: The follow-up to Greta's Purpose is a story of a chocolate colored dog who, after losing faith, finds belief, a family, and a name.

Gus and Phil Stories Audio CDs:

"In the Beginning" is a story that reminds us of the power of true friendship.

"The Baseball Toss" shows our young listeners the importance of developing Christian values.

"The Courage Contest" shows how bravery and showing off are put in their place when faith enters the picture.

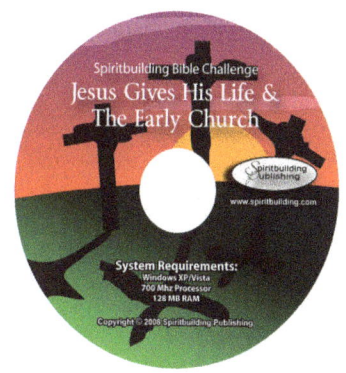   

Spiritbuilding Bible Challenge CDs

- Over 4,000 questions covering the Old and New Testaments
- For use in Bible classes, lesson reviews, supplemental class work, Bible labs, homeschooling and more

The Snow That Just Wouldn't Stop! follows the first story in the Gus & Phil audio CD word for word so it's ideal for young readers. It examines the simple pleasures of friendship and mischief in a family setting.

Bucky Beaver is a beautifully illustrated children's book, with Bible verses throughout, and is a delightful book that teaches biblical lessons of obedience and diligence while working with a cheerful heart.

www.ingramcontent.com/pod-product-compliance
Lightning Source LLC
Chambersburg PA
CBHW081024040426
42444CB00014B/3339